Nirmal Mazumder
Ankur Gogoi

Nano-structured silica on diatoms and its optical properties

AF168233

Nirmal Mazumder
Ankur Gogoi

Nano-structured silica on diatoms and its optical properties

LAP LAMBERT Academic Publishing

Impressum / Imprint

Bibliografische Information der Deutschen Nationalbibliothek: Die Deutsche Nationalbibliothek verzeichnet diese Publikation in der Deutschen Nationalbibliografie; detaillierte bibliografische Daten sind im Internet über http://dnb.d-nb.de abrufbar.

Alle in diesem Buch genannten Marken und Produktnamen unterliegen warenzeichen-, marken- oder patentrechtlichem Schutz bzw. sind Warenzeichen oder eingetragene Warenzeichen der jeweiligen Inhaber. Die Wiedergabe von Marken, Produktnamen, Gebrauchsnamen, Handelsnamen, Warenbezeichnungen u.s.w. in diesem Werk berechtigt auch ohne besondere Kennzeichnung nicht zu der Annahme, dass solche Namen im Sinne der Warenzeichen- und Markenschutzgesetzgebung als frei zu betrachten wären und daher von jedermann benutzt werden dürften.

Bibliographic information published by the Deutsche Nationalbibliothek: The Deutsche Nationalbibliothek lists this publication in the Deutsche Nationalbibliografie; detailed bibliographic data are available in the Internet at http://dnb.d-nb.de.

Any brand names and product names mentioned in this book are subject to trademark, brand or patent protection and are trademarks or registered trademarks of their respective holders. The use of brand names, product names, common names, trade names, product descriptions etc. even without a particular marking in this work is in no way to be construed to mean that such names may be regarded as unrestricted in respect of trademark and brand protection legislation and could thus be used by anyone.

Coverbild / Cover image: www.ingimage.com

Verlag / Publisher:
LAP LAMBERT Academic Publishing
ist ein Imprint der / is a trademark of
OmniScriptum GmbH & Co. KG
Heinrich-Böcking-Str. 6-8, 66121 Saarbrücken, Deutschland / Germany
Email: info@lap-publishing.com

Herstellung: siehe letzte Seite /
Printed at: see last page
ISBN: 978-3-659-71371-2

Zugl. / Approved by: Assam, India,Tezpur University, 2009

Copyright © 2015 OmniScriptum GmbH & Co. KG
Alle Rechte vorbehalten. / All rights reserved. Saarbrücken 2015

ACKNOWLEDGEMENTS

I sincerely express my gratitude to my advisor, Dr. Amarjyoti Choudhury, my major professor, who provided me tremendous support and patience in helping me to complete this project. His generous advice helped me to success both in my academic field and my life.

I would like to thank Dr. Alak Buragohin, my co-guide of this project. He always generously shared his brilliant ideas with me and offered many helpful suggestions to me.

I would like to thank Mr.Ankur Gogoi, research scholar of department of physics, Tezpur University, for his kind passion in every moment of doing this work. Again, I would like to thank Mr. Ranjan Dutta Kalita, research scholar of department of molecular biology and biotechnology, Tezpur University, for helping me in culturing diatoms and setting up different instrument. I would like to thank both of them for their help and encouragement whenever I met a problem.

During my project, I got accessed to many facilities in Department of Molecular Biology and Biotechnology and Department of Chemical Sciences of Tezpur University.

It was my great experience working in this field. I thanks to all of you. I will be very grateful if I can continue the research in my future study.

Table of Contents

Chapter 4:Optical properties of diatoms

Chapter 4

Chapter 1

1.1.Introduction:

Nanostructured silicon oxides semiconductor materials have attracted much attention because of their optoelectronic properties, particularly photoluminescence (PL) in UV- visible range. These materials have potential applications in the field of fabrication of optoelectronic devices. Nature has an ability to synthesize nanostructure composite materials that possess unique optical properties. The formation of inorganic minerals under the control of an organism (biomineralization) is a widespread phenomenon in nature. A few classes of organisms fabricate metal oxides with nanoscale features by a bottom-up self assembly process.[1] In particular, diatoms are prolific class of single celled micro-algae which can fabricate diverse 3D silica pore arrays called frustules with diameter ranges from nanometer to micrometer dimension. Diatoms are living in aquatic habitats with huge amounts, which is the bio-inspiration source of inorganic materials. The main aim of this report was to investigate the optical properties mainly focused on photoluminescence (PL) properties of diatom frustules, which grow in laboratory technique. In this report, it is found that culture of photosynthetic freshwater diatoms some are identified as *cycoltela sp.* and some were unidentified, in silicon rich environment possess optoelectronic properties, including strong blue photoluminescence were presented in the photoluminescence spectra. In recent years, these nanostructure frustules hold promises to biometric nanofabrication and devices.[2]

The characterization of the diatom materials that were cultivated by laboratory technique is accomplished in this report. In this report, the main aim is to study the luminescence properties of unique nano-porous biosilica material for the development of solar cell. Silicon, has gained a lot of interest for making widespread use in electrical, optoelectronic and structural materials. The frustules of nanometer-scale of almost 200000 species have found to be new advanced materials due to their structural and optical functions (38). Porous materials find utility in a wide variety of commercial applications. Microporous materials, such as zeolites with pores less than 1nm, have tremendous internal surface area and consequently are widely used as molecular sieves, selective sorbents as well as highly active shape –selective catalysts in the petroleum industry (2). Interest in porous semiconductor and insulation materials developed from the realization that porous silicon luminescence efficiently in the visible region when irradiated with ultraviolet light (3). The luminescence properties of porous silica have been investigated because of their relevance to the question of the origin of the luminescence of porous silicon-the silica layer being present on the surface of porous silicon. The light emission and optical properties of porous silica are of interest in their own right because of the potential to develop novel photonic devices for optical fiber-based technology (8).The development of porous silicon between 1nm and 100nm is of commercial interest as they are compatible with IC fabrication, and their ability to be integrated into microelectronic silicon based technology will allow production of low cost optoelectronic components and systems where pores can confine electrons and alter the optical emission energies (3).

There are recent reviews on diatom molecular biology, biotechnology, biomimetics, and frustule formation. This progress report focuses on the developments and potentials for diatom frustules as advanced materials by examining frustule structural features, [39] biomimetic synthesis of novel silica-based materials, chemical transformations, and templating techniques. It is reported that porous frustules act as bio-template for fabrication of gold nanostructures films. Possible applications including new nanofabrication techniques, chemo and biosensing,

5

particle sorting, and control of particles in micro- and nanofluidics will be highlighted (38).

According to a team of U.S. researchers who successfully decoded the genome of a particular diatom named *Thalassiosira pseudonana*, these very small algae could become the next big breakthrough in computer chips. They also could be used to remove carbon dioxide from the atmosphere to reduce the effects of the global climate change occurring right now.

1.2.Diatoms:

1.2.1. Definition and classes

Diatoms are one of the most beautiful organisms (~ 1-200µm in length) to look at under a microscope! They are unicellular and eukaryotic microorganisms that form an important component of the aquatic ecosystem. Diatoms are protists belonging to phylum Bacillariophyta and class Bacillariophyceae. Their cell walls are shaped like tiny glass pillboxes, with an amazing array of sizes, shapes and ornamentation. Diatoms show enormous species diversity and are inhibit both in fresh and marine water environments. There are more than 200000 diatoms species, and each of them forms unique frustules (cell wall) made of amorphous silica composites.

1.2.2.General morphology of the diatom frustule

Diatoms are single celled micro-algae that possess silica shells or frustules with intricate submicron scale features, including 3D pore arrays. Diatoms are generally classified as one of two groups depending upon symmetry of their frustules. These may be elongate with a bilateral plane of symmetry or they may be round and racially symmetrical. Many diatoms are slightly asymmetrical though they generally fall into one of these two categories.

The skeleton of a diatom or frustule is made up of very pure amorphous silica coated with a layer of organic material. This skeleton is divided into almost equal halves that fit together like a Petri dish, one of which (epitheca) overlaps the other (the hypotheca) like lid of a box joined with the connecting bands. The connecting bands together form a gridle, which are also made of silica. During diatom replication, the two halves of the frustule separate and new valves and girdles are synthesized intracellularly within specialized organelles (silica-deposition vesicles, or SDVs). During frustule development, membrane-bound transporters take actively up the soluble silicon in the form of Si(OH)4, silicic acid (5).

There are two basic body shapes of diatoms based on symmetry, centric (round with radial symmetry, Order Centrales) and pennate (thin ellipse with bilateral symmetry, Order Pennales) and either may be found in the plankton or on the benthos. While the pennate diatoms are solitary cells (although often living in dense assemblages or even forming tubes together), the centric diatoms may be solitary or chain forming, linked by projections from their cell wall or membrane. These two major taxonomic divisions also reflect a major ecological difference. Centric diatoms are mainly holoplanktonic or meroplanktonic, with only a few genera that are associated with substrates throughout their life cycles. Araphidineae (pinnate diatoms without a raphe system) and Monoraphidineae (pennate diatoms with one raphe system on one valve) are attached to sand grains, rocks and biological substrates, whereas genera of the Biraphidineae (pennate diatoms with two raphe systems on both valves) are almost completely attached to mud with only a few planktonic species. They are the major primary producers (both pelagic and benthic environment) and are thought to be responsible for up to 25% of the world's net primary productivity (Jeffery and Hallegraeff 1990) whereas total phytoplankton account for up to 40% of the global primary production (Falkowski 1994). At least some can live heterotrophically in the dark, if supplied with a suitable source of carbon (Round et al. 1990). Dinoflagellates form the other important component of the marine and freshwater phytoplankton.

1.3. Diatoms in nanotechnology

Recently there are great deal of interest centered on the design and manufacture of devices of nanometer proportions that will create a new industry termed nanotechnology. Diatoms are microscopic, single-celled algae that possess rigid micro to nanostructure cell walls (frustules) composed of amorphous silica. Depending on the species of diatom and the growth conditions, these frustules can display a wide range of different morphologies. It is possible to design and produce specific frustule morphologies that have potential applications in nanotechnology. Diatoms are touted as paradigm for the bio-fabrication of nanostructured silica (19), and many applications of the emerging fields of "Diatom nanotechnology" and "Silicon biotechnology" have been proposed (20) due to their capability of formation of 3D nanoscale architecture of silica shell.

The removal of organic materials from diatoms leaves a single valve, a solid precipitate of silica preserving the detail seen in the valve of the living cell. It has been reported that other materials might be incorporated into this structure and leading to new nanostructures.

Diatoms are interesting for nanotechnologists because of the nanoscale architecture of their silica shell, which exceeds the capabilities of present day human engineering. Nanopositioning is one of the most intriguing challenges in nanoelectronics and photonics. As the optoelectronic circuit components become more and more complex, the importance of nanopositioning becomes critical (14).

Nanoindentation is one of the most important experimental techniques to probe the mechanical properties of extremely small volumes of material. With the help of this technique the hardness and Young's modulus and hardness varied over a broad range. The observed variation in the mechanical properties is attributed to the high porosity, non uniform distribution of porosity, variations in pore size and shapes, orientation of the frustule etc (14).

The silica which forms the diatom frustule is an extremely convenient material. They are largely amorphous, (15) although the potential that extremely small crystalline zone occurs within the silica. This enables the diatoms to cast their cell material in to any from within the silica deposition vesicles. Structures constructed with such materials could thus be deposited without any negative effects on the

8

environment. So, nanotechnological devices made from silica would contribute on the environmental sustainability of anthropogenic activities (15).

Diatoms produce diverse three dimensional structures that due to their exponential rate of growth may be used in the manufacture of components for nanotechnology as an alternative to present linear lithographic techniques (16).

Nanoengineering of new materials that leads to conversion of Diatoms into 3D Structures with new Chemistries. Diatom frustules inspire the design and production of novel nanostructured materials and have recently led to new nanomaterials through the maintenance of the diatom nanostructure with modification of the material chemistry (42, 43, and 44). These nanomaterials and methods are summarized in Figure 3. The frustule of the diatoms with their various geometries and pore sizes offer a wide selection of attributes that can be exploited in nanoengineering due to their unic $_3$e nanopore structure, micro channels, chemical inertness, and silica microcrystal structure suggest many nanoscale applications such as heterogenous catalysis and separation technology (38).

The application being explored is magnetization of frustules by researchers, but there are several unanswered questions. Diatoms can be cultured in an iron-rich environment, and iron particles are doped into the nanostructured frustules. A possible method for the incorporation of magnetic particles, such as ferrite, would use standard solution chemistry techniques. As the porous frustules are known to take up aqueous solutions (18), Fe salts dissolved in water may be introduced into the porous structure. Iron oxide particles within the pores can then be produced by a suitable oxidizing heat treatment. Magnetizing frustules have many potential engineering and medical applications. Medicines/vaccines can be leaded into the pores, and then delivered by magnetic field manipulation at a target location in a human/animal body. Embedding diatom frustule in a metal-film membrane, magnetizing frustules for pin point drug delivery and producing silica nanopowders(17).

The optical properties of nanostructured (noble) metals show great promise for use in nanophotonic applications. When such nanostructures are illuminated with visible to near-infrared light, the excitation of collective oscillations of conduction

9

electrons – called surface plasmons – generates strong optical resonances. Moreover, surface plasmons are capable of capturing, guiding, and focusing electromagnetic energy in deep-sub wavelength length-scales, i.e. smaller than the diffraction limit of the light. This is unlike conventional dielectric optical waveguides, which are limited by the wavelength of the light, and which therefore cannot be scaled down to tens of nanometers, which is the dimension of the components on today's nanoelectronic ICs.

Plasmonic technology, today still in an experimental stage, has the potential to be used in future an application such as nanoscale optical interconnects for high performance computer chips, extremely sensitive (bio) molecular sensors, and highly efficient thin-film solar cells.

1.4. Diatoms as living photonic crystals

Photonic crystals are materials with spatially ordered and periodic nanostructures that can control the propagation of light, only allowing certain wavelengths to pass through the crystal similar to the propagation of electrons in semiconductor crystal (46). Photons have many advantages over electrons as carriers of information. They are faster and can convey huge amounts of data with low power losses. Photonic crystals have the potential to steer light in the same way as electrons are manipulated in semiconductor chips. They are able to control photons, producing remarkable effects that are impossible with conventional optics. The photonic crystals properties of diatoms' girdle band structures were recently confirmed (45, 31) suggesting that diatoms are living photonic crystals.

The silica cell-wall can be regarded as a photonic crystal slab waveguide with moderate refractive- index contrast. In a cell two different patterns are found: a hexagonal array of pores with a large lattice constant in the valve, and a square array of holes with a small lattice constant in the gridle. It can be demonstrated that light can be coupled into the waveguide and that there are some photonic resonances in the visible spectral range, which can be determined by band structure (3). Photonic crystal structures have been investigated in recent years with respect to their fundamental properties as well as for possible applications. The propagation of light as well as the interaction of light and matter is modified due to the presence of lattice-periodic photonic modes with characteristic dispersion relations. Applications comprise passive waveguides, diffraction elements for solar cells and active devices like photonic crystal laser.

The characteristic feature of diatoms is the biomineralised cell wall (frustule) consisting of amorphous silica. Each cell consists of two halves, the thecae, which can be divided into a valve and one or more girdle bands. The thecae overlap like a Petri dish and separate during cell division. In the cell wall of valves and girdles, very regular arrays of chambers and pores form periodic patterns, which will be the focus of our discussion. The formation of the patterns is believed to proceed via self-organized phase separation. Some kind of hierarchical order is present, leading to the self-similarity of the patterns down to small length scales. In the context, only the patterns of the order of optical wavelengths will be considered. From an optical point of view, the cell can be regarded as a 'photonic box' with walls consisting of photonic crystals (3).

1.5. Economic importance of diatoms

❖ Diatoms constitute major plankton of sea. They are the major source of food for sea animals.

❖ Due to the deposition of diatom shells in sea bed over million of years caused the formation of thousands of meter thick diatomaceous earth. Diatomaceous earth is highly absorbent and fire resistant. It is used in filters in sugar industries and brewing industries, in packing corrosive chemicals, in insulation of pipes and furnaces.

❖ It is also used in dynamite manufacture, insulation of refrigerators.

❖ It is used in polishing metals and in manufacture of toothpaste.

❖ The diatom shells can be converted to magnesium oxide, titanium oxide and BaTiO3 which are the most important semiconductor, ferroelectric materials.

1.6. Objectives

The present study of diatoms carried out with following objectives-

- ❖ Collection of environmental samples from local fresh water bodies.
- ❖ Preparation of culture media and culture of diatoms.
- ❖ Extraction of bio-silica from cultured diatoms.
- ❖ Microscopic study.
- ❖ To study the luminescence properties of bio-silica.
- ❖ FTIR analysis of biosilica.

Chapter 2

2.1. Materials and methods:

2.1.1 Equipment and chemicals:

a) Cover glass

b) Slides

c) Micropipette

d) Compound microscope with 10x, 45x and 90x magnifications.

e) Scanning electron microscope.

f) Centrifuge.

g) P^H meter.

h) Digital camera.

Chemicals:

a) Saffranine.

b) Phosphate saline buffer.

c) Nutrients for Culture media.

2.2. Methods:

Cell collection

The living fresh water diatoms are collected from the catchment ponds of a water purification plant in Assam, India during September – November, 2008.

Microscopic examination:

     The samples revealing the presence of diatom like structure is taken by compound microscope. The samples were centrifuged to 9000 rpm for 10 minutes. The heavy particles were allowed to sediment. The precipitate was suspended in 1 ml of PBS buffer to remove the artifacts to improve resolutions. The palate was again dissolved in distilled water. Then the samples are ready for culture both in liquid and solid media.

2.3.Culture of diatoms

     We can speed up evolution specially for microorganisms in the lab, using mutagenic chemicals and assorted methods for selecting those mutants that will do best for these conditions. Here the collected fresh water diatom species were cultured in the Tissue Culture Laboratory, in WC medium of Guillard and Lorenzen (1972) [12]. The protocol was slightly modified by doubling the composition of sodium meta silicate [table1] which is the source of silicon for making nanostructured silica and lowering the P^H from 7 to 6.23 hence making the culture media acidic. Cultivation of diatom cells under controlled delivery of soluble silicon offers a means to make micro to nanostructured silica. The freshwater diatoms are cultured both in the solid and liquid media.

     For the liquid culture all the growth nutrients are dissolved in 1000ml of sterile water and the media was autoclaved at 120^0c for 20 minutes. After autoclave vitamins were added in same proportions to media [table 1] and then inoculated with the environmental samples containing the micro-algae.

     For the solid media, agar is taken (1-2 % made up in water or culture medium). The inoculum is streaked or spread thinly on the agar plates. Both the liquid and solid culture is then maintained in the BOD incubator under controlled conditions of temperature $25+-0.5^0C/$ $20+-0.5^0C$ day/night cycles; photoperiods 16 hr light (fluorescent lamps) and 8 hr dark period. At the beginning the growth of the culture remains at the lag phase for 14 days after culture. For 21-30 days it is on the stationary phase. After the stationary phase the growth phase declines. During this culture process, cell numbers density increases with time and soluble silicon concentration decreases due to uptake of silicon by diatoms.

2.4. Composition for culture media:

For pH = 6.23

Fresh water "WC' medium (Guillard and Lorenzen, 1972).

Table1: Major nutrients and micronutrients for freshwater "WC" medium.

Compositions	Amount(mg/L)
Major nutrients (in mM)	
$CaCl_2.2H_2O$	36.76
$MgSO_4.7H_2O$	39.97
$NaHCO_3$	12.60
K_2HPO_4	8.71
$NaNO_3$	85.01
$Na_2SiO_3.9H_2O$	56.84
Micronutrients (in μM)	
Na_2EDTA	4.36
$FeCl_3.6H_2O$	3.15
$CuSO_4.5H_2O$	0.01
$ZnSO_4.7H_2O$	0.022
$CaCL_2.6H_2O$	0.01
$MnCL_2.4H_2O$	0.18
$Na_2MoO_4.2H_2O$	0.006

H₃BO₃	1.0

Vitamins:

Thiamin HCL – 0.1 mg/L

Biotin   -- 0.5 μ g/L

B_{12} -- 0.5 μ g/L

2.5. Frustules Isolation

Most structure in the diatom frustules are so fine that by using Light Microscopy (LM) and Scanning Electron Microscopy (SEM) can be achieved. The organic components of cell must therefore, be removed. Many methods have been developed to do this (Hassle 1978, Krammer and Lange-Bertlot 2000) some of these have some advantages and disadvantages. Acid treatment is a common method for removing all organic matrixes of the cell.

In order to analyze the diatom frustules by scanning electron microscopy (SEM) and photoluminescence (PL) spectroscopy, a cleaning procedure was needed that removed the external organic matrix covering the frustules. In the work, this was done by using the following procedure:

(a) Culture flask was shaken for 5 minutes to detach all diatoms;

(b) 10 ml of sample was centrifuged at 5000 rpm for another 10 min;

(c) The pellet was washed in double distilled water three times to remove the excess of fixative;

(d) 37% aqueous HCl was added and centrifuged at 3000 rpm for 10 minutes and was put in water bath for 15 min at 60^0C;

(e) The acid was pipetted and pellet was washed again in double distilled water 3 times. Cleaned frustules valves were then stored in ethanol to avoid contamination and bacteria growth.

Chapter 3

Microscopic study

3.1.Optical or Light microscopy (LM)

The optical microsope is a type of microscope which uses visible light and a system of lenses to magnify images of small samples. It is the oldest and simplest of the microscope. However, new designs of digital microscopes are now available which use a CCD camera to examine a sample and the image is shown directly on a computer screen without the need for expensive optics such as eye-pieces. Other microoscopic methods which do not use visible light include scanning electron microscopy (SEM) and transmission electron microscopy (TEM). Light microscopic analysis of living diatoms, in this report was performed using ZIESS.

Preparation of slides for microscopic observation of diatoms

- ❖ Slides were cleaned with detergent soap and ethanol before used.
- ❖ Take 10ml of diatom samples containing the culture media and centrifuge with 4000 rpm to 10 min and supernatant were pipetted.
- ❖ The pellet was washed with distill water.
- ❖ Then put this on the glass slide and cover with safranin stain and adjust the eye-piece to get the image.

Instrument

Basic components of optical transmission microscope are-

1. Eyepiece.
2. Objective turret.
3. Objective lenses.
4. Coarse adjustment knob.
5. Fine adjustment knob.

18

6. Object holder or stage.

7. Mirror or light (illuminator).

8. Diaphragm and condenser.

Principle

The optical components of a modern microscope are very complex and for a microscope to work well, the whole optical path has to be very accurately set up and controlled. Despite this, the basic optical principles of a microscope are quite simple.

The objective lens is, at its simplest, a very high powered magnifying glass *i.e.* a lens with a very short focal length. This is brought very close to the specimen being examined so that the light from the specimen comes to a focus about 160 mm inside the microscope tube. This creates an enlarged image of the subject. This image is inverted and can be seen by removing the eyepiece and placing a piece of tracing paper over the end of the tube. By carefully focusing a brightly lit specimen, a highly enlarged image can be seen. It is this real image that is viewed by the eyepiece lens that provides further enlargement.

In most microscopes, the eyepiece is a compound lens, with one component lens near the front and one near the back of the eyepiece tube. This forms an air-separated couplet. In many designs, the virtual image comes to a focus between the two lenses of the eyepiece, the first lens bringing the real image to a focus and the second lens enabling the eye to focus on the virtual image.

In all microscopes the image is viewed with the eyes focused at infinity (mind that the position of the eye in the above figure is determined by the eye's focus). Headaches and tired eyes after using a microscope are usually signs that the eye is being forced to focus at a close distance rather than at infinity.

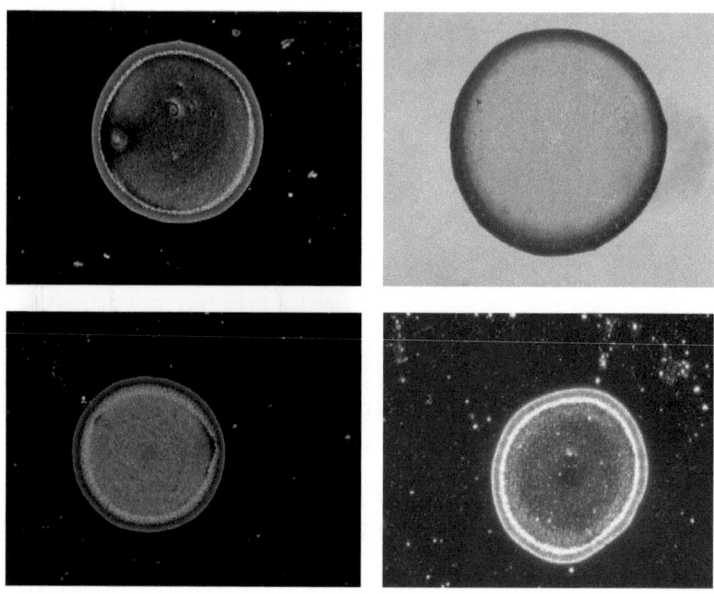

Fig.4. Optical Micrograph of diatoms with magnification **X 400** when organic matrixes are not cleared.

Living cells that were collected from freshwater bodies were observed under light microscope were shown in figure. The change of their images occurred when they were cultured in laboratory technique in Fig. The living cells were dispersed in coverslips and were allowed to settle down. The cells were oriented onto the coverslips immediately after landing, one valve of most cells was in contact with the substrate and other was facing upward. Mostly middle of the living cells covered by a thickly organic matrixes which shows in Fig., therefore, middle of the cells are not clearly observed but in cell walls some gaps are shown, which are very distinct when it was observed under Scanning Electron Microscope (SEM). Some figures of diatoms are shown with organic matrixes by optical microscope with magnification X100, X400, and X1000. It was found that the diatoms are centric. The pores were not identified but some structure of cell wall is seen which is clear in SEM image.

3.2 Scanning Electron Microscopy (SEM):

Optical microscope uses light for magnification of sample has limited resolution which can not detect very small objects. But scanning electron microscope uses electron bean which is shorter wavelength than photon to observe the matter on the surface of the sample with atomic resolution. Scanning electron microscopy has been applied to the surface studies of metals, ceramics, polymers, composites and biological materials for both topography as well as compositional analysis.

Principle: In this technique, a high energy electron beam is generated by the microscope is focused onto the sample surface that kept in a vacuum by electromagnetic lenses since electron has dual nature of character of particle as well as wave an electron beam can be focused like an ordinary light. The beam is then scanned over the surface of sample. The electron beams are scattered i.e. how many number of electrons are scattered back and forth from the sample and fall on to the detector and then to a cathode ray tube through an amplifier. The signals that derive from electron-sample interactions reveal information about the sample including external morphology (texture), differences of atomic number within the sample or information about the elemental composition, crystalline structure and orientation of materials making up the sample. The SEM is also capable of performing analyses of selected point locations on the sample; this approach is especially useful in qualitatively or semi-quantitatively determining elemental compositions (using EDS), crystalline structure, and crystal orientations (using EBSD).

The accelerated electrons in an SEM carry significant amounts of kinetic energy, and this energy is dissipated as a variety of signals produced by electron-sample interactions when the incident electrons are decelerated in the solid sample. These signals include secondary electrons, that produce SEM images, are then amplified ,analyzed and translated for showing morphology and topography on samples, backscattered electrons(BSE) are most valuable for illustrating contrasts in composition in multiphase samples, diffracted backscattered electrons (EBSD) that are used to determine crystal structures and orientations of minerals, X-ray generation is produced by inelastic collisions of the incident electrons with electrons in discrete orbitals (shells) of atoms in the sample, which will give the elemental composition of the sample.

3.2.1. Sample preparation of frustules for SEM analysis

The purpose of SEM measurements is to observe the surface topography of diatom frustules, measure the size of the pores on the surface of the diatom skeleton. The frustules of three different species of diatoms were isolated by hydrochloric acid treatment and analyzed by JEOL JSM 6390 LV Scanning Electron Microscopy (SEM) and X-ray energy dispersive analysis (EDS) probe. For SEM analysis, about 20 µL of frustule suspension in methanol were pipetted onto a rectangular coverslip and then dried in oven at 60°-70° C for two to three days. The coverslips containing the dried frustules were mounted onto SEM stubs with carbon sticky tabs. An electrical conduction bridge were formed between the stubs and samples by coating the edges with platinum in a JEOL JFC 1600 auto fine coater and viewed through SEM.

Fig.6: SEM image of *Cyclotela sp*. of freshwater diatom at low magnification.

Fig.7: SEM image of *cyclotella cryptica sp*. of freshwater diatom.

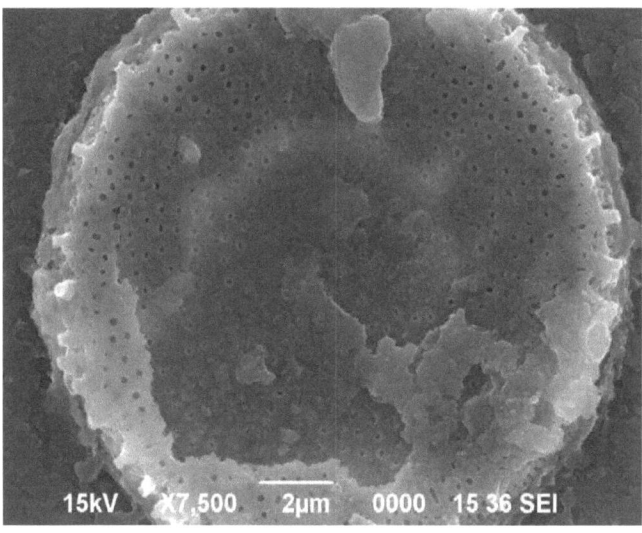

Fig. 8: SEM image of unidentified sp. of freshwater diatom.

Fig. 9: SEM image of unidentified sp. of freshwater diatom.

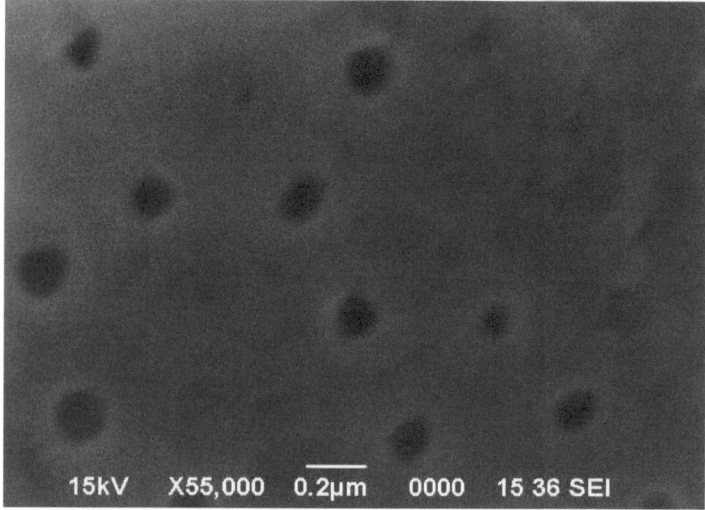

Fig. 10: is a high magnification SEM image of the diatom frustule showing pores are as an average size about 100nm in diameter.

Scanning Electron microscopy (SEM) images of representative frustules obtained by hydrochloric acid treatment of different diatoms cell biomass from the control silicon rich cultivation at stationary state are presented in figure(6,7,8,9,10). From SEM images it was confirmed that the aqueous hydrochloric acid treatment procedure did not always break the frustules, but still gives distinct images. SEM images clearly show the surface structure of the cell wall of diatoms as circular, cylindrical identify. After all the soluble silicon was depleted from the medium, the surface of the diatom biosilica become granular at the nanoscale (figures 3 and 4) clearly showing that frustules were composed of uniform distributions of nanopores. The sizes of diatoms are in micrometer ranges. The average diameters of nanopores are about 100 nm

3.2.2. Energy Dispersive X-ray Spectroscopy (EDS or EDX)

Energy dispersive X-ray spectroscopy (EDS or EDXRF) is the standard procedure for identifying and quantifying elemental composition of sample area as small as a few cubic micrometers (13). X-ray generation is produced by inelastic collisions of the incident electrons with electrons in discrete orbitals (shells) of atoms in the sample. The detection of these X-rays can be accomplished by an energy dispersive spectrometer, which is a solid state device that discriminates among X-rays energies. At rest, an atom within the sample is in ground state i.e. unexcited electrons in discrete energy levels or electron shells bound to the nucleus. As the excited electrons return to lower energy states, they yield X-rays that are of a fixed wavelength (that is related to the difference in energy levels of electrons in different shells for a given element). Thus, characteristic X-rays are produced for each element in a mineral that is "excited" by the electron beam. SEM analysis is considered to be "non-destructive"; that is, x-rays generated by electron interactions do not lead to volume loss of the sample, so it is possible to analyze the same materials repeatedly.

The SEM instrument used here for X-ray analyzed is energy dispersive methods. The SEM-EDS spot analysis of three different species confirmed that the frustules isolated by hydrochloric acid treatment were composed mainly of silicon in the form of SiO_2. The characteristic k-energy peaks were identified as in the tables-

Table 2.

Element	Weight (%)	Atomic (%)
O K	53.92	68.98
Al K	2.49	1.89
Si K	33.39	24.77
K K	5.31	2.78
Ti K	2.07	0.89
Zn K	2.21	0.69
Total	100.0	

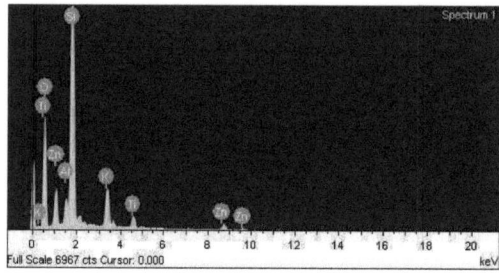

Fig. 11. EDX spectrum of *cyclotela cryptica sp.* when organic materials were removed.

Table3.

Element	Weight (%)	Atomic (%)
O K	53.70	68.91
Al K	2.52	1.92
Si K	33.74	24.67
K K	5.22	2.79
Ti K	2.21	0.95
Zn K	2.61	0.82
Total	100.0	

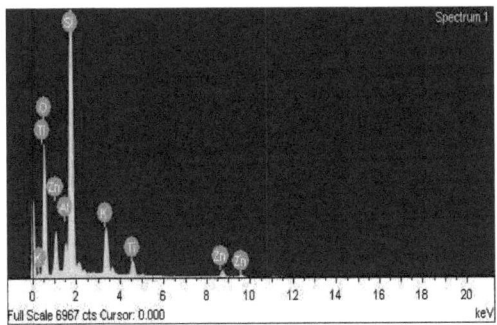

Fig.12. EDX spectrum of pinnate diatoms, image was not shown.

Table 4.

Element	Weight (%)	Atomic (%)
O K	57.77	71.59
Al K	1.12	0.92
Si K	36.72	25.92
Ni K	1.34	0.45
Ti K	1.74	0.72
Zn K	1.31	0.40
Total	100.0	

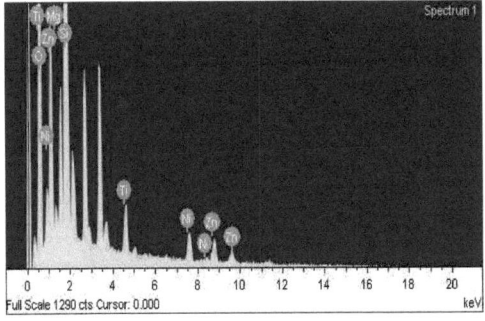

Fig.13. EDX spectrum of *cyclotela sp.* when organic materials were removed.

The corresponding SEM-EDS spectra for each species of diatoms show approximately the same results. But out of this the percentage of silicon is more in centric diatoms, *cyclotela sp.* All data were collected at an accelerating voltage of 20kV. The characteristic k-energy peaks were

The quantitative elemental analysis of nanocluster showed that 97-98 % of elemental constitute of white powder were Si in the form of SiO2. The other elemental constituents are Mg, Ca, Zn, Ti, Ni, Al and K. The atomic composition of oxygen was sufficient to ensure that the metals were in oxide forms. The trace elements in the analysis Ca, Zn, Ni and Mg were the result of nutrient uptake by the living diatoms. In EDS spectra the origination of Ti is of more important for study.

Chapter 4

Optical properties of diatoms

4.1. UV-vis absorption spectroscopy

Absorption spectroscopy uses the range of electromagnetic spectra in which a substance absorbs.

Ultraviolet-visible spectroscopy (uv = 200-400 nm, visible = 400-800 nm) corresponds to electronic excitations between the energy levels that correspond to the molecular orbitals of the systems.

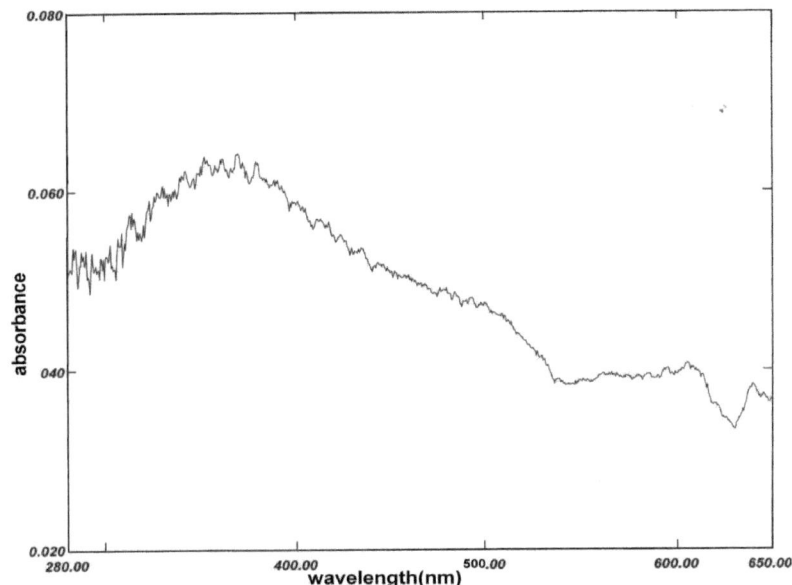

Fig.14. UV-vis absorption spectrum of frustules of diatom.

The UV-vis absorbance spectrum of the frustule of diatom is shown as in the Fig., shows a characteristic absorption peak near at 370nm. This absorption is considered to be caused by an interaction between light and inner micro to nanostructure of frustules of diatom (9) as shown in figures of SEM.

4.2. Photoluminescence spectroscopy

Photoluminescence spectroscopy technique is by far most widely used optical characterization method in porous Si investigations. A wide range of important sample properties such as electronic band structure, electron photon coupling and nature of defect and impurity centre, can be studied by PL.

The photoluminescence (PL) measurements were performed at room temperature using the PERKINS ELMER LS 55 fluorescence spectroscopy. In this report, PL measurement of both diatoms and isolated biosilica were analyzed at different excitation wavelengths and at room temperature.

The energy levels in semiconductor materials are also be investigated by using the technique of photoluminescence. A laser is used to photo excite the electrons in biosilica. The atoms are spontaneously emitting luminescence when electrons in excited states drop down to a lower level by radiative transitions. The luminescence is analyzed with a spectrometer and the peaks in the spectra represent a direct measure of energy levels in the semiconductor.

Principle:

If a light particle (photon) has energy greater than the band gap energy, then it can be absorbed and thereby raise an electron from the valance band to conduction band across the forbidden energy gap. In this process of photo-excitation, the electron generally has excess which it loses before coming to rest at the lowest level of conduction band. At this point electron eventually falls back down to the valance band. As it falls down, the energy it loses is converted into luminescent photon which is emitted from the materials. Thus the energy of limited photon is a direct measure of band gap energy, E_g. The process of photon excitation followed by photon emission is called photoluminescence.

Photoluminescence (PL) is the fundamental property for any optoelectronic devices. Luminescence from porous silicon and Si nanocrystallites is very sensitive to surface structure of crystallites. The surface effect plays an important role other than passivation. There are several possibilities for origin of luminescence from porous silicon. (a) the pure quantum confinement effect in silicon nanostructure.(b) siloxane-based molecular compound. (c) silicon hybride complexes. (d) polysilanes. (e) surface localizations states on Si-nanocrystallites (38).

Some researchers studied PL property of synthesized silica, and they observed the similar blue PL peak for bulk crystalline and amorphous SiO2. It was reported in Tanimura et al (12) that for semiconductors, the recombining donors and acceptors are separated by a distance which is a few times of lattice constant, while the luminescence band in a-SiO2 is considered to arise from self-trapped exciton, in which the electron-hole separation is of the order of the nearest O-O separation in the lattice. Thus these characteristics of recombination of a-SiO2 can be explained on the basis of self-trapped exciton model. They have concluded their experimental results that the luminescence band around 2.1 eV in a-SiO2 arises due to self-trapped exciton. Some researchers also pointed out that the STE PL is peaked between 2.6 eV and 2.8eV for crystalline SiO2, whereas it is extended over the entire visible spectral range with a maximum between 2.1 and 2.4 eV for amorphous silica. The difference in peak positions of the STE PL reflects the features of the self-trapped excition formation crystalline and amorphous SiO2. The photoluminescence of nanostructure silicon can be explained by the radiative recombination of the electron-hole pairs generated in a quantum-confined system.

The quantum confinement (QC) effect in semiconductor nanoscale materials (nanoparticle, nanoclusters, quantum dots, boxes, wires etc.) has recently received considerable attention for its potential use in optoelectronic devices (10). The luminescence is often interpreted as resulting from quantum confinement in small size crystallites of porous silicon. The band-gap energy increases with decreasing size of nano-objects, this effect may account for an observed shift in the excitonic photoluminescence (PL) towards higher energies in comparison with bulk materials and consequently enhances quantum efficiency (11). Thus it is customary to explain any blue shift in the excitonic PL with decreasing nano-object size as resulting from the quantum confinement effect. Yuri D. Glinka et al (10) two models were applied to examine the blue shift of STE PL band with decreasing size of nanometer-sized silica fragments. The model based on quantum confinement effect on Mott-Wannier type excitons developed for semiconductor nanoscale materials. The model is completely unusable in the case of wide-band-gap nanosale materials (the band-gap of bulk silica E_g=11 eV).

The PL measurements reported here are related to samples obtained from three different species of diatoms that cultured in laboratory. The diatom cells were treated with hydrochloric acid to study the photoluminescence properties of diatom cell. The visible PL effect from diatom is clearly seen after espousing silica structure to UV light with a broad blue luminescence peak in the visible range centered at 450nm. This effect is similar to PL of artificially fabricated porous Si. This PL is strongly species dependent and it is based on both their frustule structure and surrounding environment. At room temperature, three different species show very similar spectra. The origin of visible PL of diatom frustules is still unclear, on the other hand, it is known that metallic impurities are commonly incorporated inside the amorphous silica frustules (in particular Zn, Mg and Cu) which was confirmed by SEM-EDS analysis and visible PL emission may also arise from metal-oxygen complexes whose PL can be significantly quenched by adsorbed via energy transfer reactions (7). Consequently, the photoluminescence emitted from diatom should be closely related to the nano-sized amorphous silica. The blue PL emission peak likely most originated from surface defect sites associated with the nanostructured biosilica, particularly the nanoparticle assembles constituting the solid biosilica.

The photoluminescence spectra from frustules of diatoms (Fig.16, 17, 18) consist of one broad asymmetric peak centered at around 440nm (2.8 eV). All the photoluminescence peaks are at 2.7 eV, thus the major reason caused the blue photoluminescence (PL) from diatoms could be self-trapped excitons (STE) and quantum confinement effect, because the frustules of diatoms are consisted of nano-sized $SiO2$.

Quantum confinement effect is used to describe some phenomena occurred when the diameter of the particle is of the order of magnitude of the wavelength of the electron wave functions, the conductive band will be split to the covalent band. The band gap will increase with the decrease of the diameter (13). Quantum confinement describes the increase in energy which occurs when the motion of a particle is restricted in one or more dimensions by a potential well. As the confining dimension decreases, the energy of particles increases (41). A crystal nanoparticle is treated as a well that is confined in all three dimensions. Therefore, the energy bandgap of the nanocrystal is much higher than the bulk crystal reported the observation of visible photoluminescence (PL) of Si nanoparticles (40).

Thus the material shows more obvious photoelectron effects such as the quantum confinement effect, delocalization quantum coherence effect and non-linear effect (6). In that case, specially optical and electrical properties of the material have some deviations from their bulk materials. Since the frustules of diatoms are composed mainly of amorphous SiO2, so band-gap should be higher than 11eV which is band-gap of bulk SiO2. With this high band-gap electrons are hard to be excited from the valence band to conduction band. In this report, the excitation wavelength 335nm, 370nm which are much lower than the bandgap of the diatom sample. Thus, with this excitation light, the luminescence measured from the diatom samples should not be caused by band-to-band transition. The most possible reason for photoluminescence from diatoms is by self-trapped exciton under quantum confinement effect.

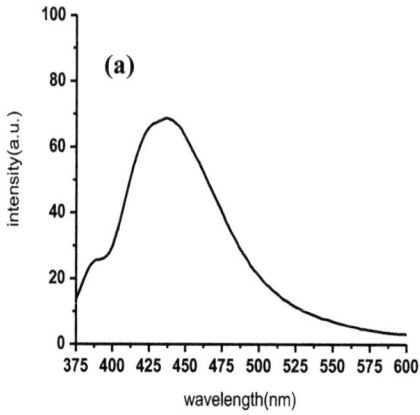

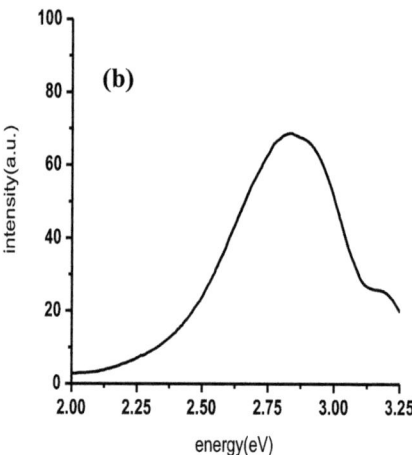

Fig.15. Photoluminescence (PL) spectra of frustules at stationary stage at excitation wavelength 337nm.(a) wavelength (nm) vs intensity(a.u.). (b) energy (eV) vs intensity(a.u.).

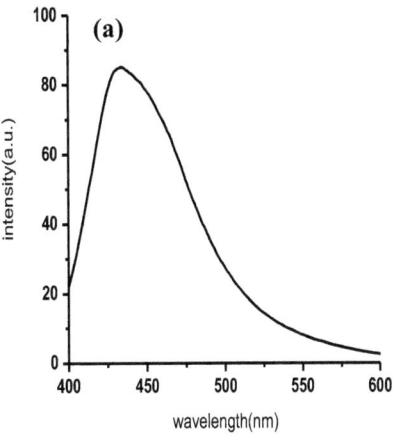

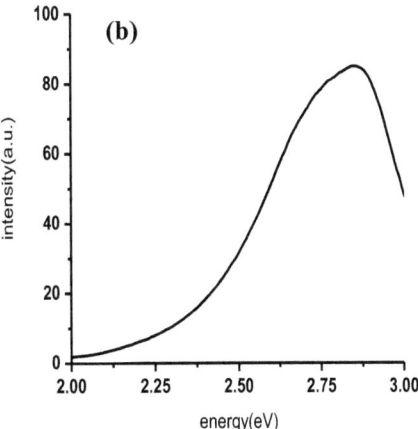

Fig.16. Photoluminescence (PL) spectra of frustules of diatoms at room temperature at stationary stage at excitation wavelength 370nm. (a) wavelength(nm) vs intensity(a.u.). (b) energy (eV) vs intensity(a.u.).

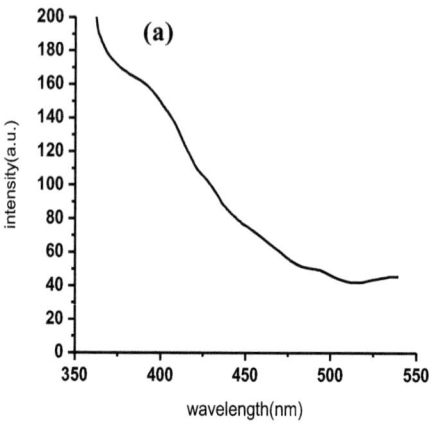

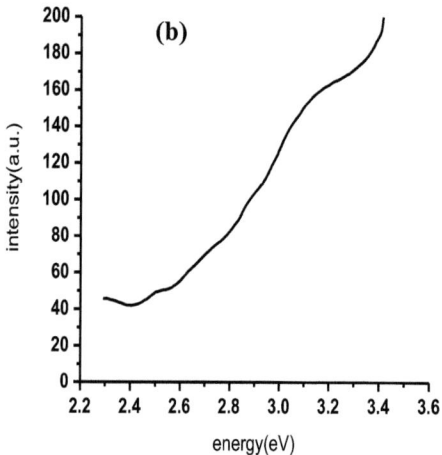

Fig. 17. Photoluminescence (PL) spectra of frustules of diatoms at room temperature at stationary stage at excitation wavelength 337nm. (a) wavelength(nm) vs intensity(a.u.). (b) energy (eV) vs intensity(a.u.).

4.3. FT-IR Spectroscopy

Fourier Transform Infrared (FT-IR) spectroscopy is a very common technique to obtain the bond information of chemical substances. When radiation passes through a sample, certain frequencies of the radiation are absorbed by the molecules of the substance leading to the molecular vibrations. The frequencies of absorbed radiation provide the characteristics of a substance. FT-IR spectrum is obtained by first collecting an interferogram of a sample signal using an interferometer, and then performing a Fourier Transform on the interferogram to obtain the spectrum. Unlike dispersive spectrometer (grating monochromator), in which wavenumbers are observed sequentially as the grating is scanned, FT-IR spectrometer collects all wavelengths simultaneously. This makes the signal collection much faster (is) than using dispersive spectrometer (30min).

FT-IR spectroscopy is increasingly being used for analysis of bio-macromolecules. By combining IR spectroscopy with modern molecular biological techniques, detailed structure-function relationships on a local level can be revealed (47).

The chemical structural analysis of the frustules of diatom at ambient temperature carried out by means of Fourier Transform Infrared Spectroscopy (FTIR). FT-IR measurements were performed at 2.0 cm−1 resolution on a Nicolet Instruments 410 FT-IR equipped with KBr optics and a DTGS detector. Frustules (2.0 mg) were mixed with 100 mg KBr and ground to fine powder, and 40 mg of this mixture was compressed for 1.0 min to form a transparent pellet which was mounted onto the FT-IR sample holder.

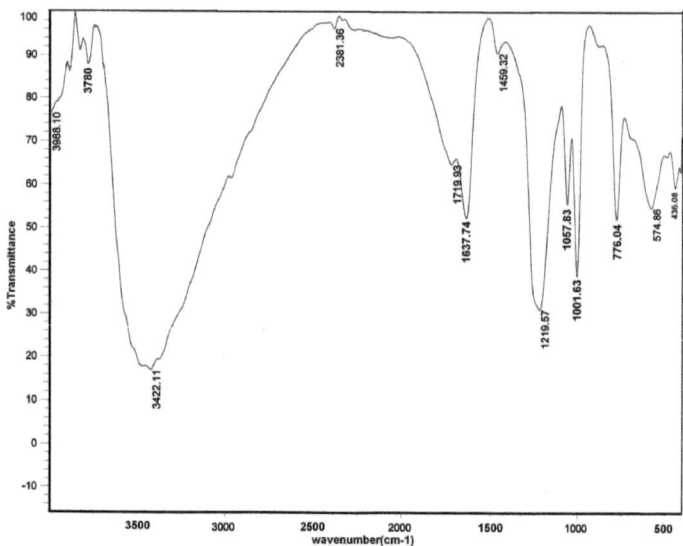

Fig.18. Infrared spectrum of diatom frustules measured in transmittance mode using FTIR spectrometer (ref. 54)

FT-IR spectra of *cycloteta sp.* frustule powder isolated by aqueous hydrochloric acid treatment of diatom cells obtained at stationary stage is presented in Figure 18. FT-IR spectrum as recorded in transmittance mode a clearly showed characteristic peaks for diatom biosilica, including for Si-O-Si stretching vibration at 1057 and 776 cm-1 (2), and O-H stretching of bound water surface hydroxyl groups at 3422 cm-1, which could also include Si-OH stretching mode. Also the characteristic peak for carboxyl (C-O) stretching of esters or fatty acids at 1735 cm-1 was obtained. Therefore, residual organic materials were still adhered to the biosilica frustule even after acid treatment. The peak at 1630 cm−1 is ambiguous and could represent carboxyl (C-O) stretching for primary amides of residual proteins associated with diatom biosilica (5), which could include the "silaffin" class of silica precipitation proteins imbedded within diatom biosilica,18 H-O-H bending of adsorbed water,32 or Si-OH stretching (2). Consequently, the frustule biosilica possessed silanol (-Si-OH) groups but not silicon hybride (Si-H) groups.

Chapter 5

5.1. Discussion

Diatom nanotechnology conceived as an industrial process, is now a highly interdisciplinary, fast moving area, possibly headed for making a major contribution to nanotechnology. Bio-nanotechnology also driven as an industrial pursuit might nevertheless make fundamental contributions to our understanding life. There are many contributions of diatoms in experimental field, such as electroluminescent device fabrication, electroluminescence measurement, photonic band calculation, dielectric constant measurement.

Although, the experiments done in these a-SiO2 were under different conditions than that in this report. The data found above provided a comparison with the results of photoluminescence from diatoms. The differences and compatibilities between the PL experiments in these papers and that in this report includes, the material in the paper is pure synthesized SiO2 nanoparticles, but in this report the material is the naturally biomineralized SiO2 nanomaterial which comprises some impurities.

PL band at 2.8 eV was observed from the biosilica. The properties of the PL bands appear to be very complex and strongly depend on the oxygen content in biosilica. However, together with the results of FTIR and at room temperature PL, it may find out a correlation between the PL properties and the nanostructure of frustules. Zhao et al (48) observed the 415-437 nm Pl band from evaporated nanocrystalilne Si and proposed that it is due to a zero dimensional quantum confinement effect (QCE) in Si quantum dots. In agreement with recent PL measurements on amorphous silicon nanostructures (49), crystallinity is a necessary requirement to observe quantum confinement effects on the electron wave function in silicon and silicon alloys (50). So, QCE as the possible origin of the 2.8 eV PL band.

Combining the FTIR data with the evoluation experiment provides useful information for understanding the 2,8 eV luminescence. Fig.18 shows that the peak of Si-O-Si bond evidenced by IR signature around 1057 cm-1 and the Si-O-H bonds with a stretching vibration around 3420 cm-1 exit in the biosilica.

From the above observations, it is reasonable to suggest that the 2,8 eV PL band is related to a defect associated with the OH groups. Although bulk SiO2 is an insulator with a wide band gap of ~11eV, defects associated with nanostructured silica give rise to visible photoluminescence. The origin of blue PL in nanostructured silica is still under debate, but surface silanol groups (-Si-OH) are the most possible source of blue PL emission. Tamura et al (51) have observed a visible PL with a maximum around 450 nm from postoxidized nanocrystalline silicon film. They proposed that this luminescence is associated with the absorption of silanol group Si-OH in the silica network. This result generally agrees with this report. Thus the blue photoluminescence of the laboratory cultured cyclotela sp. biosilica originates from surface silanol (Si-OH) groups associated the fine nanoscale features on the diatom frustule.

Other properties of nanostructured silica may also indirectly affect PL properties. For example, decreasing the diameter of the nanopores increases hydrogen bonding interactions between neighboring surface silanol groups, this in turn shifts the peak PL to a higher energy (52).

In summary, the data summarized in Figs.15 and 16 convincingly show that the blue photoluminescence which is found in nanostructure biosilica is due to defects and the silica network on which OH groups are adsorbed.

5.2.Conclusions:

In conclusion, this study illustrates the potential of using living cells to direct the biological frabrication of nano-to-microscale inorganic materials that posses strong blue photoluminescence properties. Due to silicon starvation by the cultivated species of photosynthetic centric diatoms in laboratory induced the nanostructure of the diatom frustule, which in turn blue photoluminescence to the frustule biosilica.

Diatoms research is rapidly moving from the taxonomists to nanotechnology and big business. Diatom nanobiotechnology, a new interdisciplinary area, has successfully emerged over the past several years into a dynamic research. These significant progresses of research have helped in understanding diatom structural, mechanical, genomic,optical and photonic properties and silica biomeneralization process, leading to nanofabrication and nanoengineering of new materials and devices based on diatom silica.

Finally, Si nanostructures in diatoms are complex materials that have unique optical properties. Diatoms have the potential as factories for the production of a wide range of materials that may be of great benefit to nanotechnology and micro-to-nanofabrication. Diatom frustules are biominerals with an enormous number of different structures and patterns with micro and nanoscale features that offer considerable potential for nanofabrication and nanodevices like solar cell.

5.3. Future works

Diatom nanotechnology can be moved to industrial applications. It is now a highly interdisciplinary, fast moving area, possibly headed for making a major contribution to nanotechnology.

Silicon is the most important optoelectronic materials for most of the electronic devices. Recently, it is found a simple method of converting frustules- the silica based nanostructured of diatoms into pure silicon structures with many applications. At high temperature around 2000oC, silica reduced to liquid silicon, which on cooling produces 98 percent pure metallurgical grade silicon. Further refining affords ultra-pure silicon for the electronics industry. This naturally produced silicon can be used as a solar cell due to their environmental sustainability.

AFM study of my sample is not studied in this report, it will my future work. The nanoindentaion can be studied through AFM of *cyclotela sp.*

Correlation of Raman and photoluminescence spectra of biosilica will be my future study.

In this report, the characterization and optical properties of centric diatoms were studied. This will also be studied in case of pinnate and other species of centric diatoms.

TiO_2 and GeO_2 nanostructure can be obtained by biological fabrication, using two stage photobioreactor processes, into the biosilica of diatom *Nitzchia frustulum*. This type of fabrications with other materials is the new search of nanofabrication. Bioprocess technology is a new tool to metabolically insert doped metals into the silica frustules and produce nanocomposite materials. In future, this bioprocess of formation of nanocomposite will be studied with centric diatoms as breadboards.

It will also be the new study, how the structural and optical properties of frustules of diatoms are varied with changing concentration, PH, temperature, and adding other minerals in the culture media.

5.4. References

1. M.W. Anderson et al, J. of Nanosci. and Nanotech. 5, 92-95 (2005).

2. L. De Stefano, I. Rendina, M. De Stefano, A. Bismuto and P. Maddalena Appl.phy.lett.,87, 233902(2005).

3. L.T.Canham, Appl. Phys. Lett. 57, 1046 (1990).

4. Mario De Stefano, and Luca De Stefano, J. of Nanoscience and Nanotech. Vol 5,15-24, 2005

5. Tian Qin, Timothy Gutu, Hun Jiao, Chih-Hung Chang, and Gregory L. Rorrer, J. of Nanosci. and Nanotech. Vol-8,2392-2398, 2008.

6. ZHANG Zheng-Hua et al, Chin. Phys.Lett. vol-24, No.2 (2007) 543.

7. A.Setaro, S.Lettieri, P.Maddalena, and L.De Stefano, Appl. Phys. Lett. 91, 051921 (2007).

8. K.S.A. Butcher, J.M.Ferris, M.R.Phillips, photoluminescence and cathodoluminescence studies of diatoms-nature's own nano-porous silica structures, in: John Cashion, Trevor Finlayson, David Paganin, Andrew Smith, Gordon Troup (Eds.), proceedings of the 27th A and NZ Condensed Matter and Materials Meeting, 4-7 February, Charles Sturt University, Wagga Wagga, NSW, 2003, p. 51.

9. Yamanaka et al, J. Appl. Phys. 103,074701 (2008).

10. Yuri D. Glinka et al, Phys. Rev. B, Vol-64, 085421 (2001).

11. J.P. Proot, C.Delerue, and G.Allan, Appl. Phys.Lett. 61, 1948 (1992).

12. K. Tanimura, C. Itoh, and N.Itoh, J.Phys.C 21,1869 (1988).

13. Characterizations of silicon-germanium nanocomposites fabricated by the marine diatom Nitzschia frustulum", a thesis by Liu, Shuhong, Department Chemical Engineering, Origen State University, 4-March 2005.

14. G. Subhas, S. Yao, B. Bellinger and M.R. Gertz, J. of Nano. and Nanotech.vol 5,50-56, 2005.

15. Christian E. Hamm, J. of Nano. and Nanotech. Vol 5, 108-119,2005.

16. Richard Gorden and John Parkinson, J. of Nano.and Nanotech. Vol 5, 35-40, 2005.

17. Kit Mun Wee et al, J. of Nanosci. and Nanotech. Vol-5, 88-91,2005.

18. Clayton Jeffryes et al, ACS NANO, Vol-2. No.10. 2103-2112. 2008.

19. C. Jeffryes, T. Gutu, J.Jiao and G.L. Rorrer, Matter. Sci. Eng. C. 28, 107 (2008).

20. Ryan W. Drum and Richard Gordon, TRENDS in Biotechnology, Vol.21 No.8, Augest 2003.

21. D. Losic et al, Adv. Funct. Mater. 2007, 17, 2439-2446.

22. Shannon Dudley et al, J.Am.Cerem. Soc., 89 [8] 2434-2439 (2006).

23. H. Sandage et al, Chem. Commun.,2005,651-653.

24. M.S. Haluska et al, Powder Diffraction 20(4), Dec. 2005.

25. T. Debenest et al, J.Appl Phycol, Feb.2008.

26. Jonathan C Taylar et al, African Journal of Aquatic Science 2005,30(1), 65-75.

27. D. Losic et al, Chem. Commun.,2005,4905-4907.

28. D.Losic et al, New J.Chem., 2006,30,908-914.

29. Raymond R. Unocic et al, Chem.Commun.,2004, 796-797.

30. Clayton Jeffyes et al, Adv. Matter. 2008, 20,2633-2637.

31. Helen E Townley et al, Nanotechnology 18 (2007) 295101.

32. M.J.Estes and G.Moddel, Phys. Rev. B, Vol-54,No-20,1996.

33. Toshihide Takagahara and Kyozaburo Takeda, Phys.Rev. B, Vol-46, No-23, 1992.

34. Won Chel Choi et al, Appl. Phys.Lett. 69 (22), 1996.

35. K.S.A. Butcher et al, Mater. Sci. Eng. C, 25 (2005) 658-663.

36. Tian Qin et al, ACS Nano, Vol-2,No-6,2008.

37. S.A. Grant et al. Sens. Actuators B, 69 (2000) 132.

38. D. Losic, J.G. Mitchell, and N.H. Voelcker, Adv.Matter.2009, 21,1-12.

39. C. E. Hamm, J. Nanosci. Nanotechnol. 2005, 5, 108.

40. A. Colder, F. Huisken, E.Trave, G.Ledoux, O.Guillois,C.Reynaud, H.Hofmeister and E.Pippel, Nanotechnology 15 (2004) L1-L4.

41. "Photoluminescence Properties Investigation of Germanium Inserted Biosilica Generated by Bioreactor Culture of Marine Diatom Nitzschia frustulum" A thesis by Tian Qin for the degree of Master of Science in Chemical Engineering presented on November 25, 2008.

42. M. B. Dickerson, K. H. Sandhage, R. R. Naik, Chem. Rev. 2008, 108, 4935.

43. J. Parkinson, R. Gordon, Trends Biotechnol. 1999, 17, 190.

44. R. Gordon, J. Parkinson, J. Nanosci. Nanotechnol. 2005, 5.

45. Fuhrmann, T. et al. (2004) Diatoms as living photonic crystals. Appl. Phys. B 78, 257–260

46. Hall, N. and Ozin, G. (2003) The photionic opal – the jewel in the crown of optical information processing. Chem. Commun. (Camb.) (21), 2639–2643.

47. Arthur M.A.Pistorius. Biochemical applications of FT-IR spectroscopy. Spectroscopy Europe 7/4 1995.

48. X.Zhao, O.Schoenfeld,S.Komuro,Y. Aoyagi and T.Sugano, Phys.Rev. B. 50, 18654 (1994).

49. R.B.Wehrsphon, J.N.Chazalviel, F.Ozanam, and I.Solomon, Phys. Rev. Lett. 77, 1885 (1996).

50. M.Zhu, Y.Han, R.B.Wehrspohn, C.Godet,R.Etemadi, D.Ballutaud. J. of Applied Physics. Vol-83,No- 10, 5386-5396 (1998).

51. H.Tamura, M.Ruckshloss, T.Wirschem, and S.Veprek, Appl. Phys. Lett.65, 1537 (1994).

52. C.M.Carbonaro, F.Clemente, R.Corpine, C.P. Ricci and A.Anedda, J.Phys. Chem. B. 109, 14441 (2005).

53. Ankur Gogoi, Alak K. Buragohain, Amarjyoti Choudhury, Gazi A. Ahmed. Laboratory measurements of light scattering by tropical fresh water diatoms. Journal of Quantitative Spectroscopy and Radiative Transfer (2009), doi:10.1016/j.jqsrt. 2009.03.008.

54. N. Mazumder, A.Gogoi,R. D. Kalita, G. A. Ahmed, A. K. Buragohain, A. Choudhury, "Luminescence studies of fresh water diatom frustules" Indian Journal of Physics 2010, Vol. 84, Issue 6, pp 665-669

I want morebooks!

Buy your books fast and straightforward online - at one of the world's
fastest growing online book stores! Environmentally sound due to
Print-on-Demand technologies.

Buy your books online at
www.get-morebooks.com

Kaufen Sie Ihre Bücher schnell und unkompliziert online – auf einer der am
schnellsten wachsenden Buchhandelsplattformen weltweit!
Dank Print-On-Demand umwelt- und ressourcenschonend produziert.

Bücher schneller online kaufen
www.morebooks.de

OmniScriptum Marketing DEU GmbH
Heinrich-Böcking-Str. 6-8
D - 66121 Saarbrücken
Telefax: +49 681 93 81 567-9

info@omniscriptum.com
www.omniscriptum.com

OMNIScriptum

MIX
Papier aus verantwortungsvollen Quellen
Paper from responsible sources
FSC® C105338

FSC
www.fsc.org

Printed by Books on Demand GmbH, Norderstedt / Germany